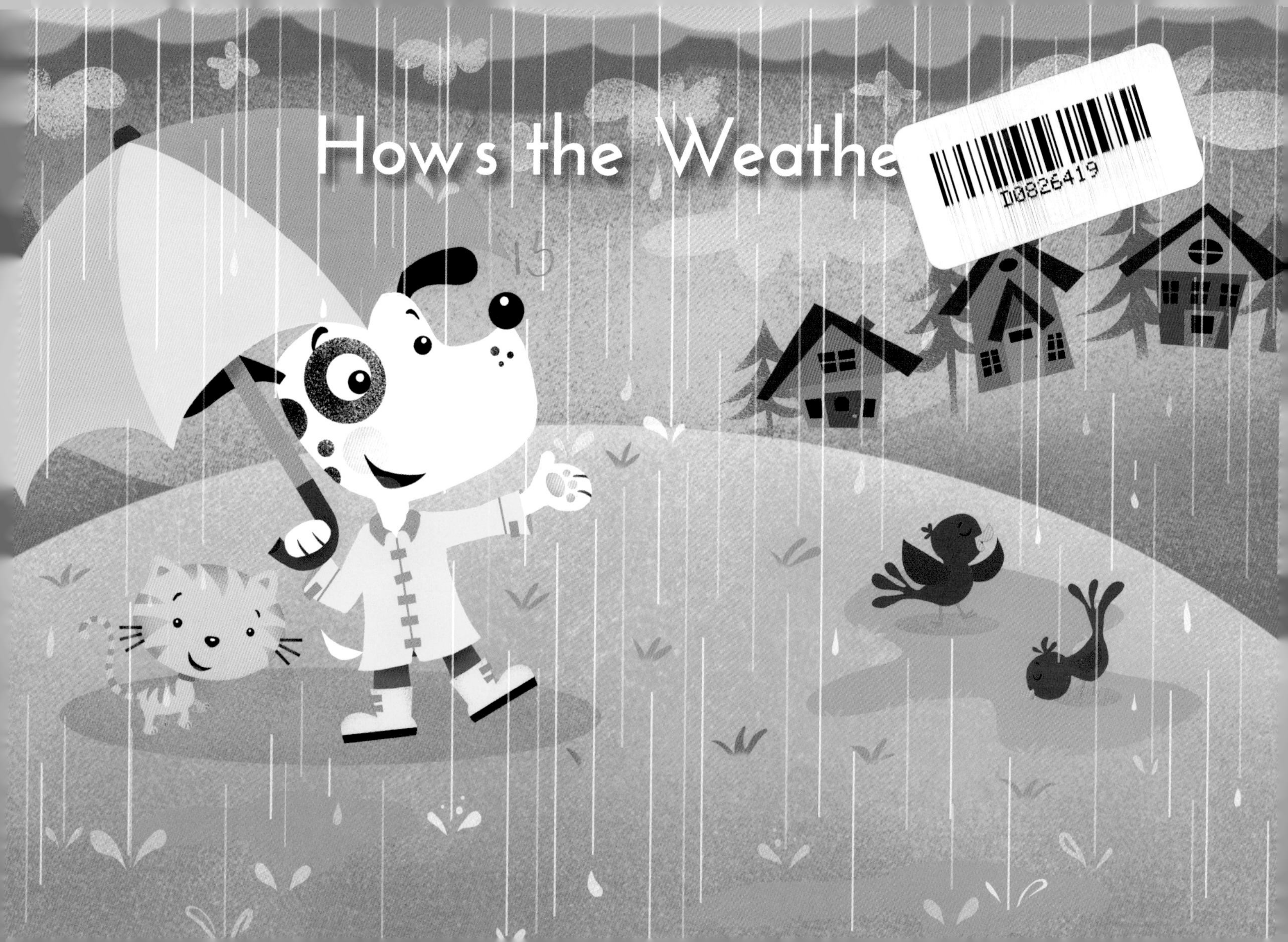
Hows the Weathe

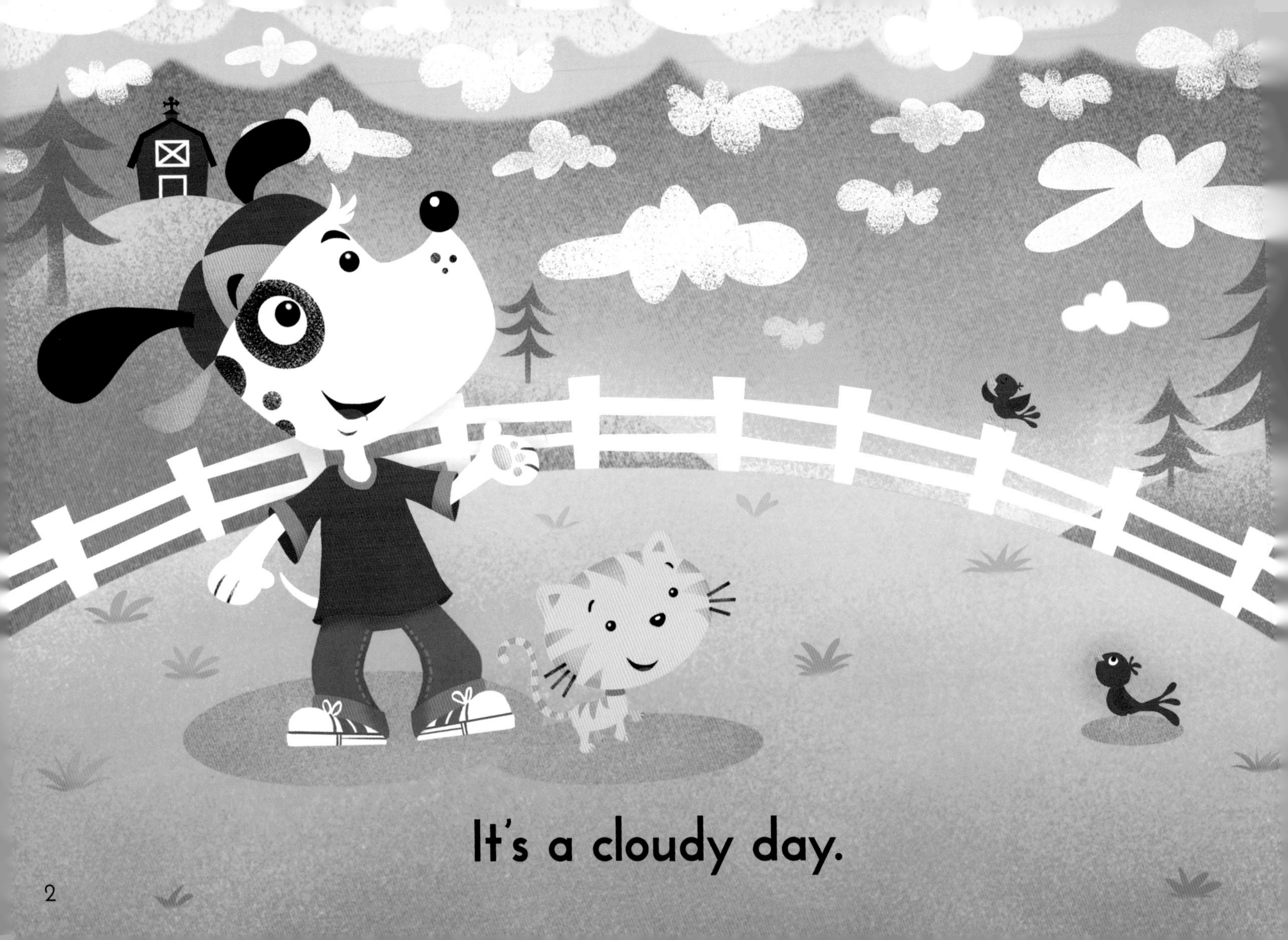

It's a cloudy day.

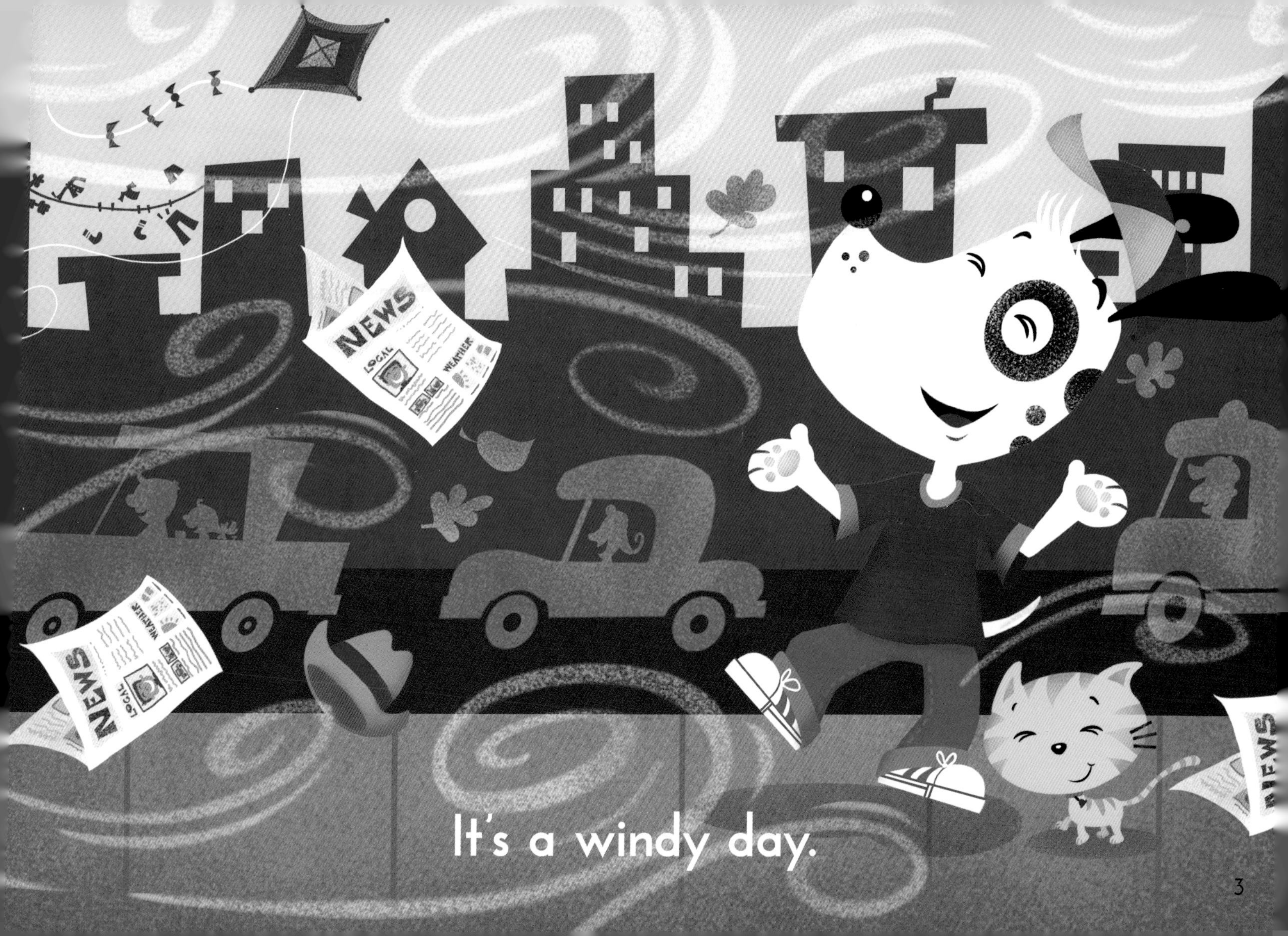

It's a windy day.

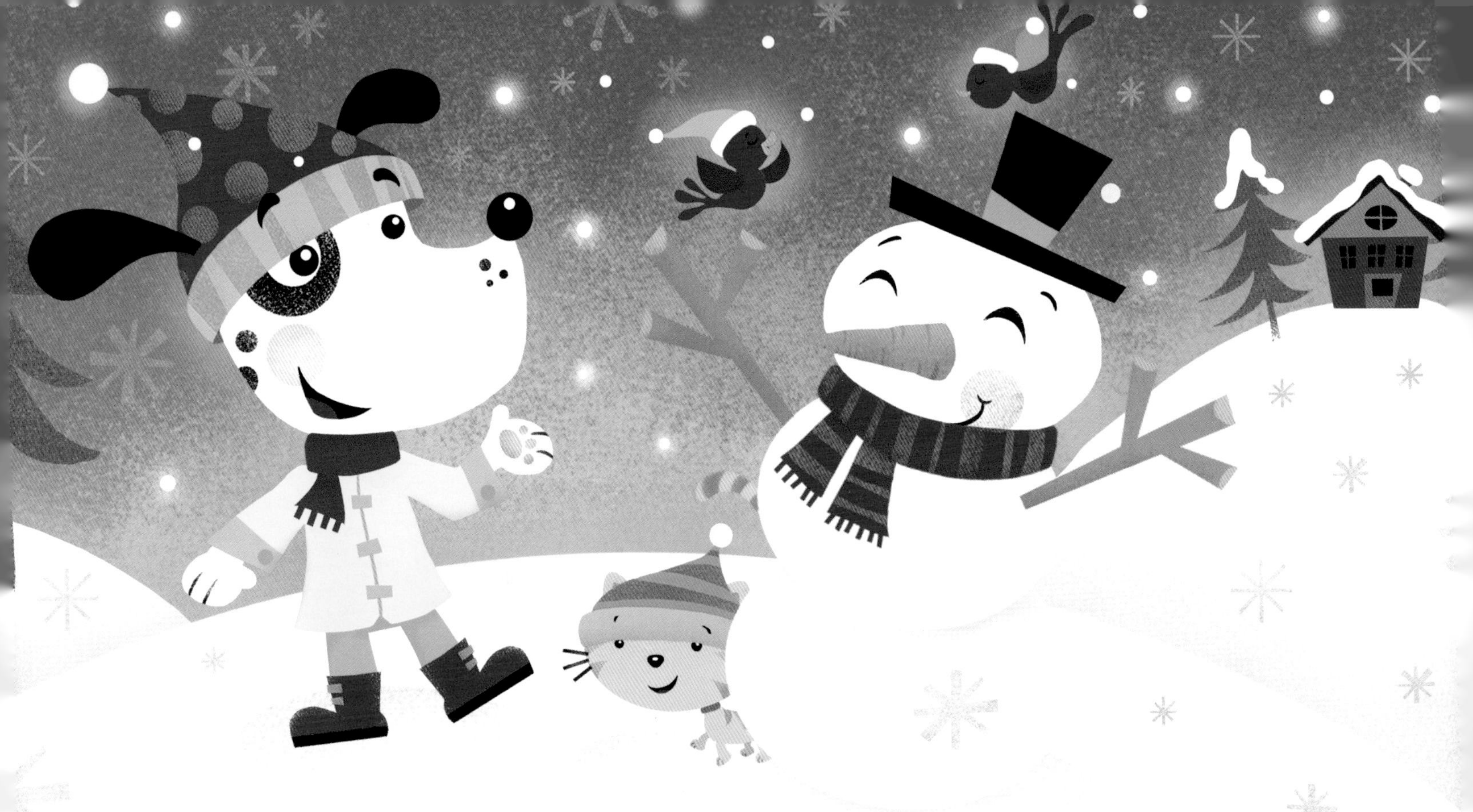

It's a snowy day.

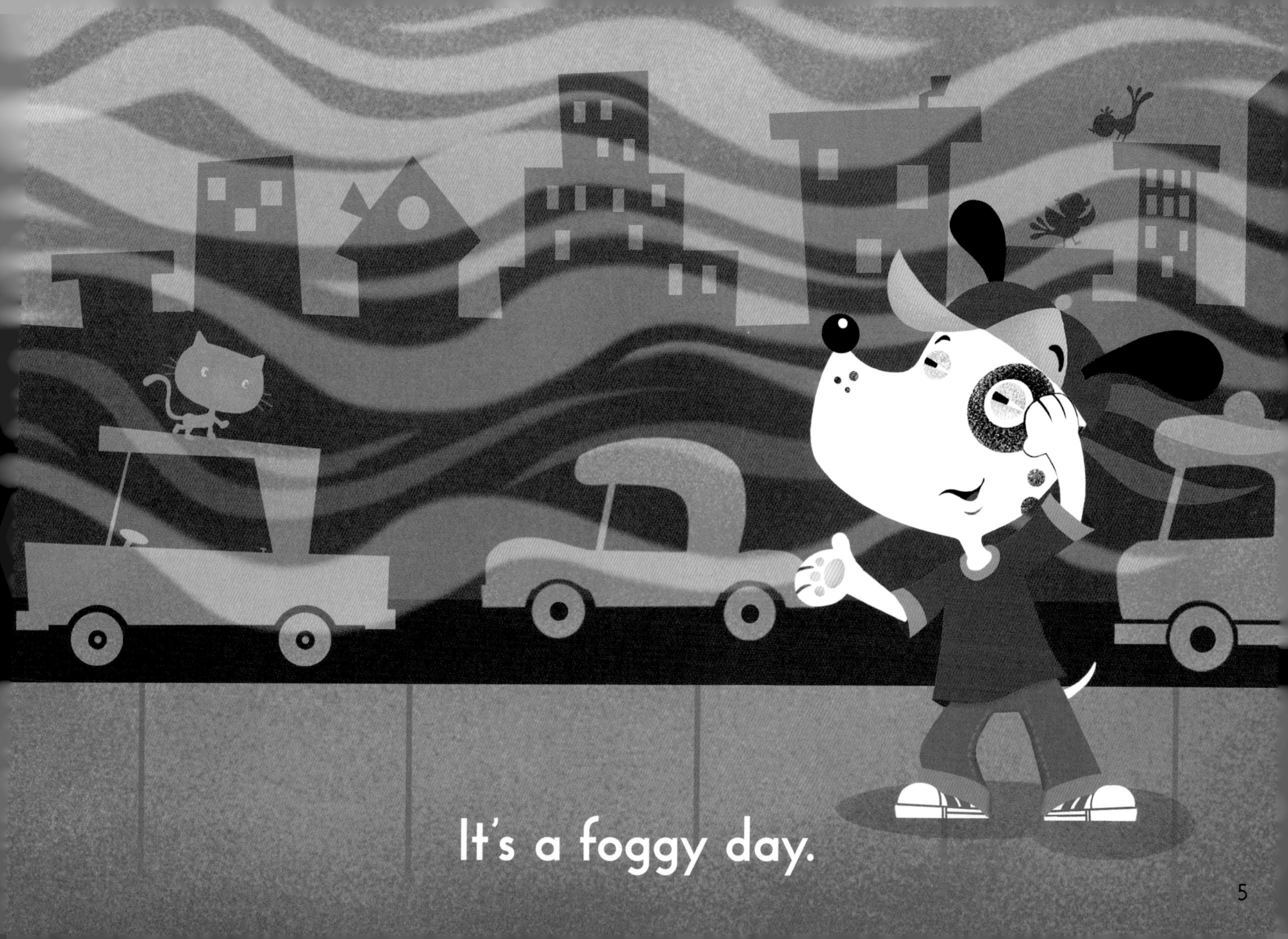

It's a foggy day.

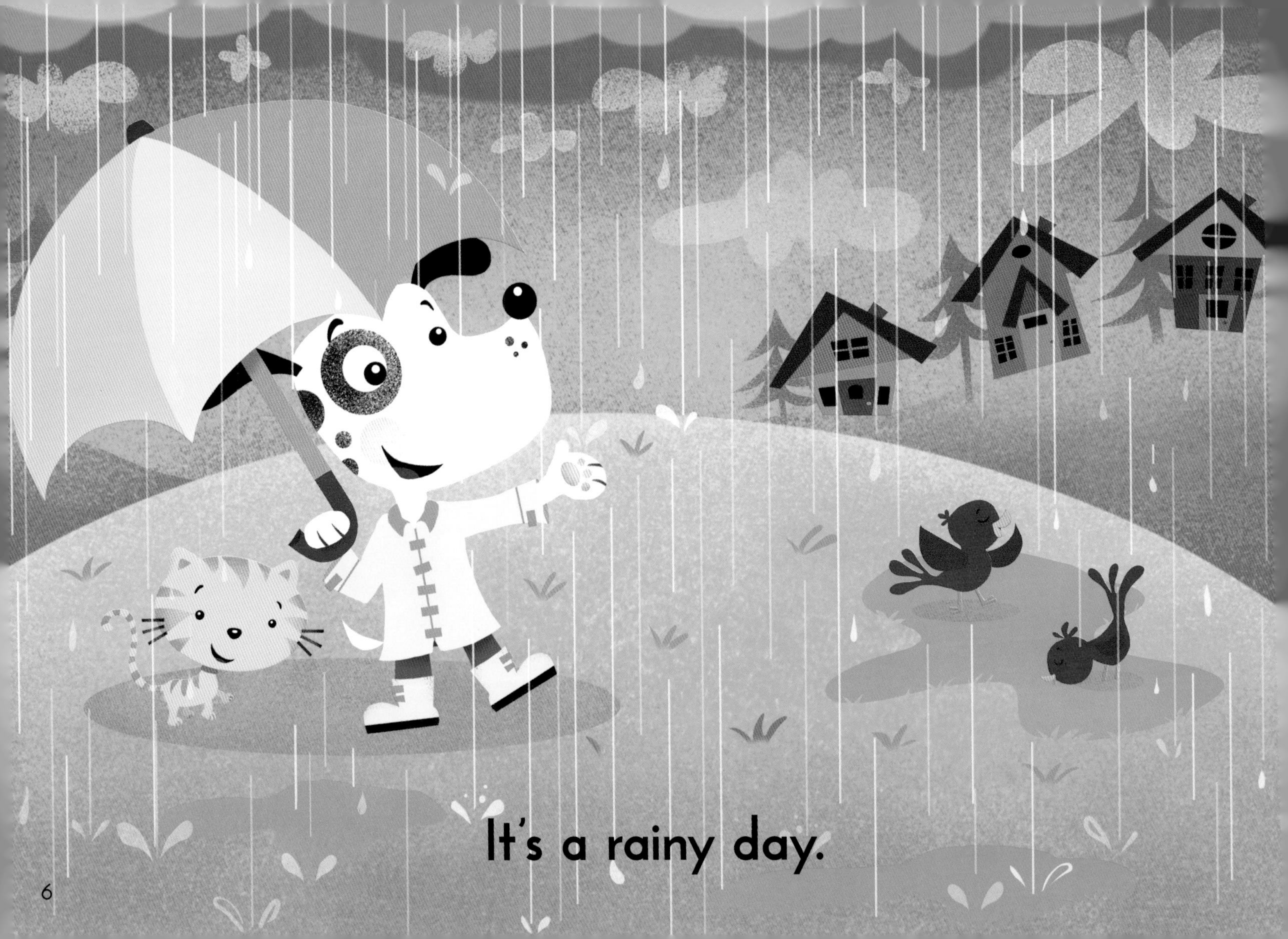

It's a rainy day.

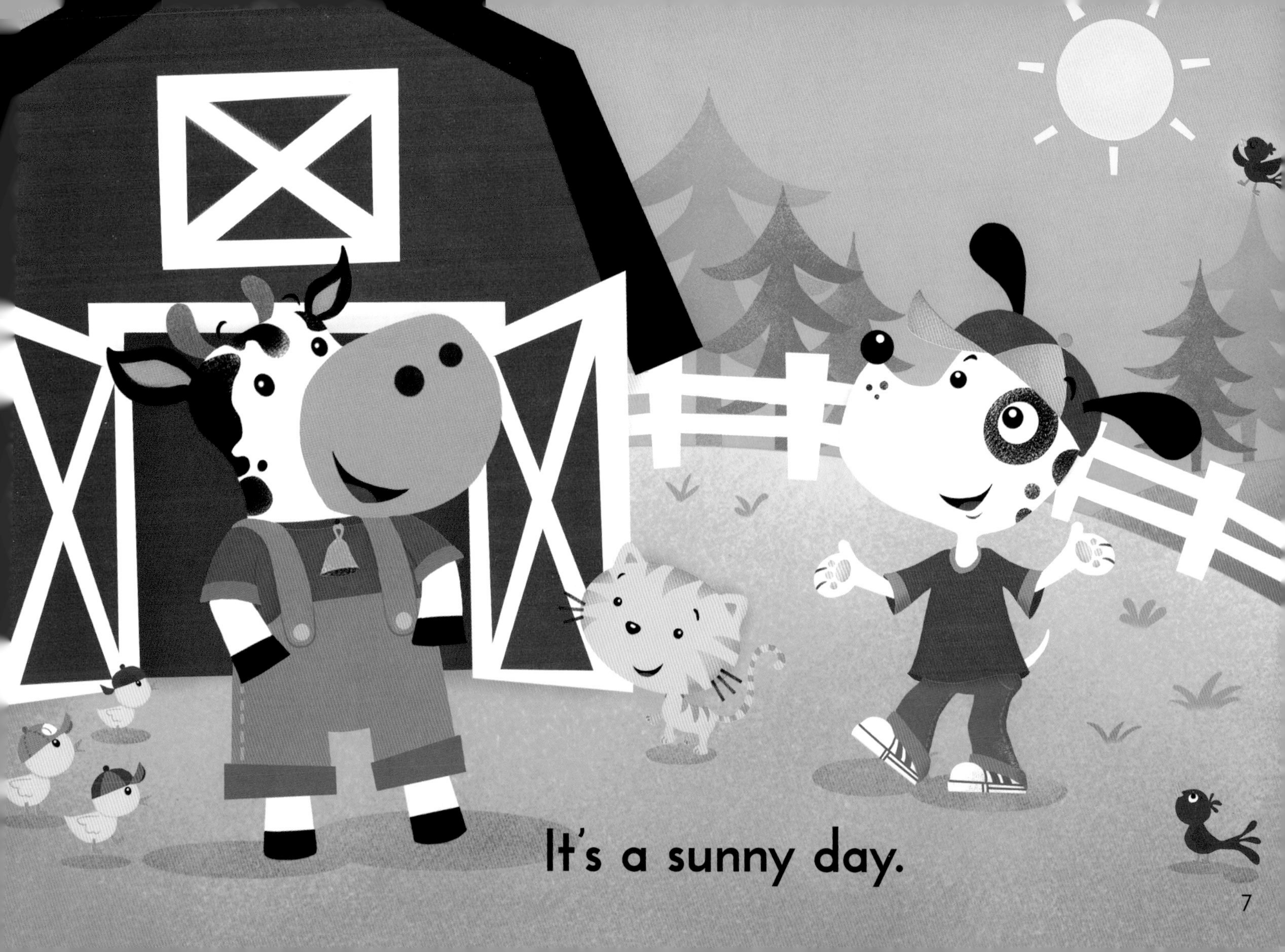

It's a sunny day.

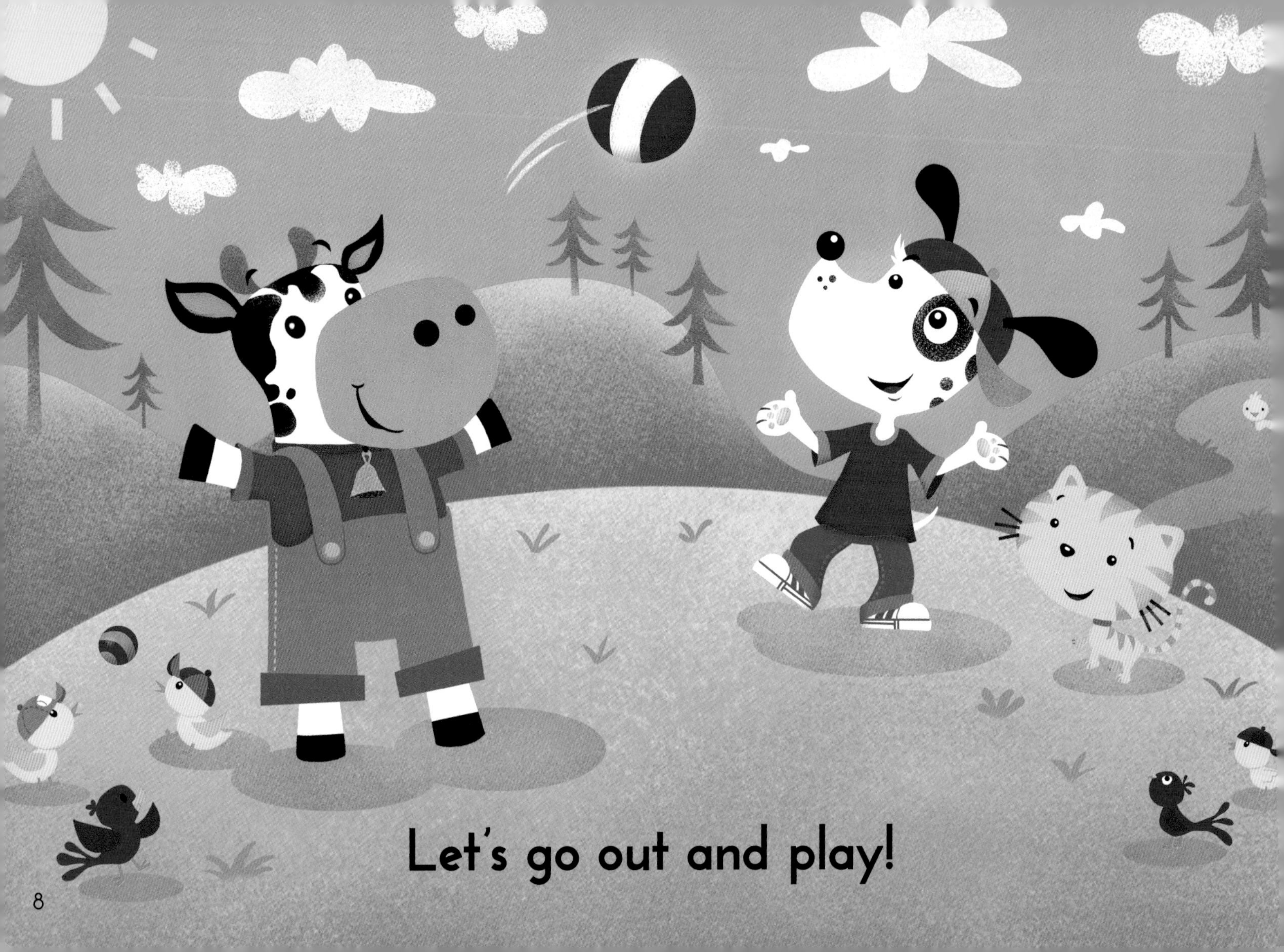

Let's go out and play!